THE HOSPITAL

Ron Thomas and Jan Stutchbury
Illustrated by Annette Dowd

Alternative Learning Materials

McDougal, Littell & Company

Evanston, Illinois

This is the hospital.

4

Crib

Many babies are born here.

6

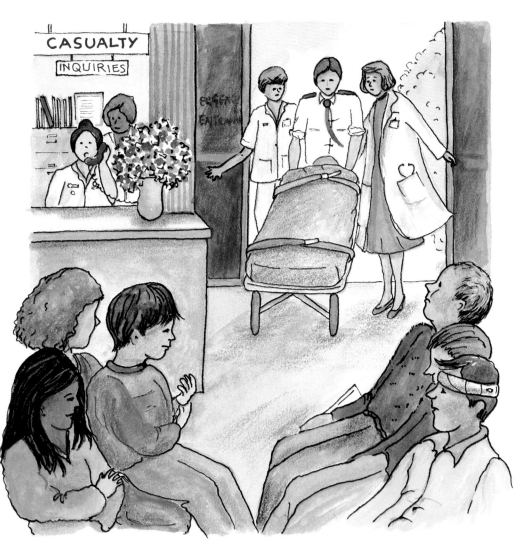

In an emergency,
when people are sick or injured,
we get help at the hospital.

Ambulance

An ambulance takes people
to the hospital.

8

Helicopter

Stretcher

There are helicopters
that are ambulances, too.
They are used for people who
are far away and who need
to get to a hospital quickly.

9

A room in the hospital
is called a ward.
Some wards have many beds,
and some have only one or two.

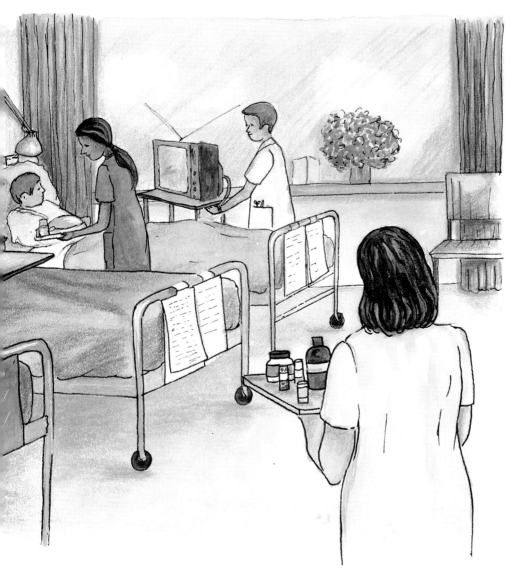

There are nurses in each ward
to look after the patients.

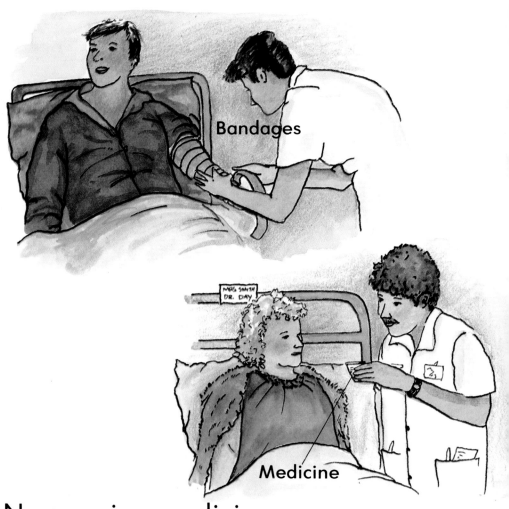

Bandages

Medicine

Nurses give medicine.
They change bandages.
They take the patient's temperature
with a thermometer.

12

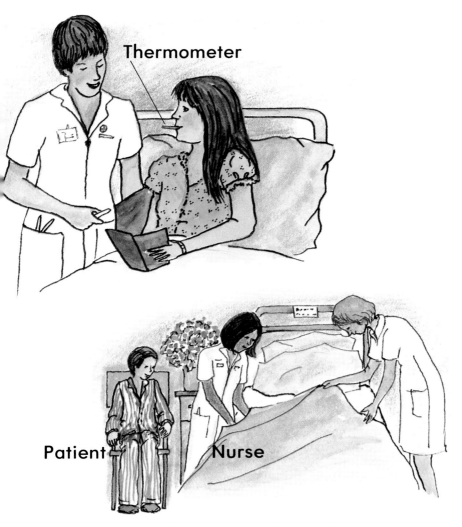

Thermometer

Patient Nurse

Nurses make sure that the patients
are getting better.
They try to make them
happy and comfortable.

13

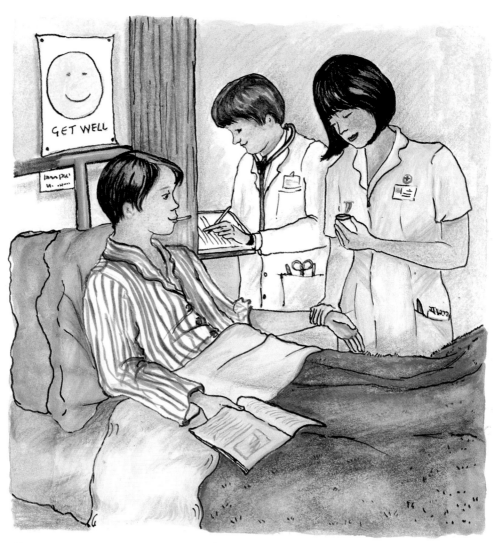

Doctors visit each patient.
The doctor says how each
patient is to be treated.

14

Prescription

Druggist

A prescription is an order
for medicine.
The doctor writes a prescription
for the patient when the patient
needs medicine.
The medicine comes from
the hospital's druggist

In the operating theater
doctors and nurses perform
an operation.

16

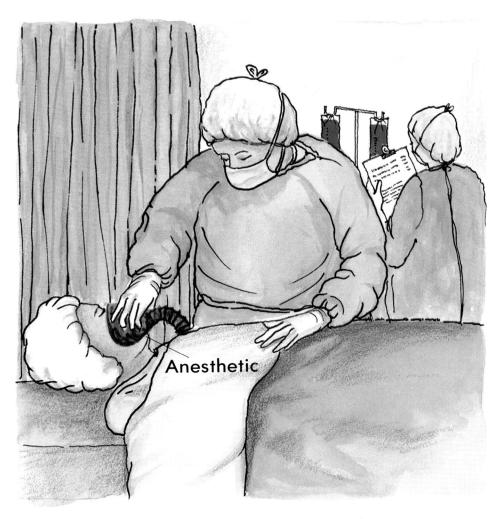

Anesthetic

Before a patient is operated on,
they are given an anesthetic.
An anesthetic puts them to sleep
so that they cannot feel anything.

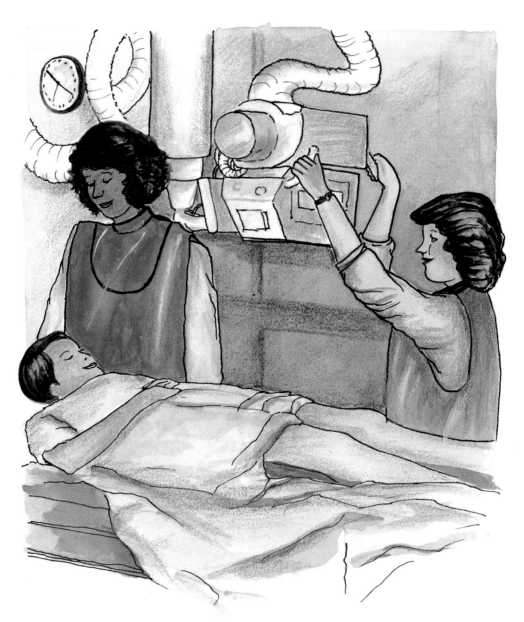

The X-ray machine takes pictures.

X-ray

The pictures show the inside
of a patient's body.

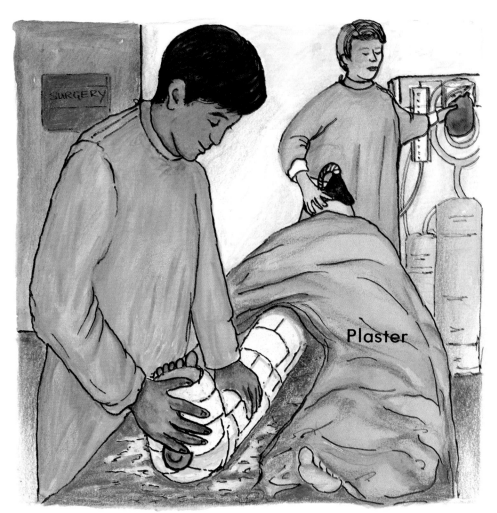

Plaster

If a bone is broken in our leg or arm plaster is put around it.
The plaster stops you from moving the bone so it can set and mend.

When the bone has mended,
the plaster is taken off.
Exercises make the patient's
arm or leg strong again.

Knitting

Patients who must stay in hospital
for a long time can learn to do many
things.

Reading

Painting

For children
there's a playroom with toys
and lots of books to read.

At the children's hospital
there are teachers to help
with school work.

24

Meals are prepared and cooked
in the hospital's kitchen.
Many people work there.

Cleaner

The hospital must be kept very clean.

26

Visitors come to cheer-up
the patients.
They can buy flowers and gifts
at the hospital shop.

The people who work in the hospital care for us and help us to get better; but it's good to go home.

Glossary

ambulance
(helicopter)

ambulance (van)

anesthetic

doctors

druggist

medicine

nurses	operating theater
patient	prescription
thermometer	visitors
ward	X-ray machine